THE AMAZING 7-DAY, SUPER-SIMPLE, SCRIPTED GUIDE TO TEACHING OR LEARNING DECIMALS

Lisa Hernandez, MS.Ed.

NOVA PRESS

Additional educational titles from Nova Press (available at novapress.net):

- **The Amazing 7-Day, Super-Simple, Scripted Guide to Teaching or Learning Fractions** (80 pages)
- **The Amazing 8-Day, Super-Simple, Scripted Guide to Teaching or Learning Percents** (90 pages)
- **GRE Math Prep Course** (528 pages)
- **GMAT Prep Course** (624 pages, includes software)
  GMAT Math Prep Course (528 pages)
  GMAT Data Sufficiency Prep Course (422 pages)
- **Master The LSAT** (608 pages, includes software and 4 official LSAT exams)
- **The MCAT Physics Book** (444 pages)
  The MCAT Biology Book (416 pages)
  The MCAT Chemistry Book (428 pages)
- **SAT Prep Course** (640 pages, includes software)
  SAT Math Prep Course (404 pages)
  SAT Critical Reading and Writing Prep Course (350 pages)
- **ACT Math Prep Course** (402 pages)
  ACT Verbal Prep Course (248 pages)
- **Scoring Strategies for the TOEFL® iBT:** (800 pages, includes audio CD)
  Speaking and Writing Strategies for the TOEFL® iBT: (394 pages, includes audio CD)
  500 Words, Phrases, and Idioms for the TOEFL® iBT: (238 pages, includes audio CD)
  Practice Tests for the TOEFL® iBT: (292 pages, includes audio CD)
  Business Idioms in America: (220 pages)
  Americanize Your Language and Emotionalize Your Speech! (210 pages)
- **Law School Basics:** A Preview of Law School and Legal Reasoning (224 pages)
- **Postal Exam Book** (276 pages)
- **Vocabulary 4000:** The 4000 Words Essential for an Educated Vocabulary (160 pages)

ISBN-13: 978–1–889057–24–8
ISBN-10: 1–889057–24–X

NOVA PRESS

9058 Lloyd Place
West Hollywood, CA 90069

Phone: 1-800-949-6175
E-mail: info@novapress.net
Website: www.novapress.net

Table of Contents

Page number	Section	Pertains to Lesson Day	Worksheets	Corresponds to Note page
4–5	Forward	Intro for Instructor		
6	Section Layout	Intro for Instructor		
7	Separate Sheet Sideways	Intro for Instructor		
8	Catchphrase Preview	Intro for Instructor		
9	The Script	Intro for Instructor		
10–13	What are Decimals?	Day 1	1	
14–16	Comparing and Ordering Decimals	Day 2	2A, 2B, 2C	A
17–18	Adding and Subtracting Decimals	Day 3	3	B
19–21	Multiplying Decimals	Day 4	4	C
22–24	Dividing Decimals	Day 5	5	D
25–28	Chunking Division	Day 6	6A, 6B	E
29–30	Multiplying and Dividing Decimals by 10's, 100's, 1000's, etc.	Day 7	7A, 7B	F
31	The Review	Day 8	8	G
33–40	The Notes	all		
41–54	The Worksheets	all		
55–66	The Answers	all		
67–70	The Appendix	all		
71–74	The Tests	Day 9		G

Additional notes, charts, and illustrated examples are available in the Appendix section

FORWARD

Dear Educator, Tutor, Paraprofessional, or Eternal Student,

Welcome to *The Amazing 7-Day Super-Simple, Scripted Guide to Teaching or Learning Decimals*. Though we know that an attempt to line up all the teacher guides ever written on teaching decimals would result in the need to colonize outer space, I have attempted to do just what the title says: make it super simple. I have also attempted to make it fun and even ear-catching. The reason for this is not that I am a frustrated stand-up comic, but because in my fourteen years of teaching the subject, I have come to realize that my jokes, even my bad ones, have a crazy way of sticking in my students' heads. And should I use a joke (even a bad one) repetitively, the associations become embedded in their brains, many times to their chagrin!

In the age of Google, there is no need for me to rant about research regarding "edutainment" or even to pretend to present a comprehensive battery of skill and drill worksheets. As teachers in this day and age, we are lucky enough that most of the information that we need on any topic or lesson plan can be found right at our keyboard or even cell phone fingertips. However, I have also found that this plethora of resources can be at times overwhelming, and at best, difficult to sort through.

Therefore, I have undertaken the mission of writing a good old-fashioned book, I have tried very hard to include no superfluous pages (as they can easily be downloaded as previously mentioned) and attack the four main operations associated with decimals, identifying each with its own catchphrase. Associations with each catchphrase (see Appendix page I) further inhibit students' inexplicable ability to completely erase fifty-minute lessons. They even manage to take students outside of their dreaded textbooks!

That being said, I consider myself an old-fashioned teacher in a not-quite-yet middle-aged body. I favor individual work over group work, have harsh opinions regarding inclusion, and rarely see a need for manipulatives. I want my students to succeed with pencil and paper, relying on themselves and not others, both of which seem to be dying arts, regardless of the necessity of these skills in the workplace. In too many urban schools, I have labored with administrators admonishing me to incorporate the newest fangled trend ("Let's just talk about our *feelings* regarding Mathematics in a big circle") to 8th graders who have never been forced to memorize their multiplication tables. If you are reading this book, you are probably old enough to realize that though piles of worksheets may not be "sexy" enough for many contemporary educational theorists, they no doubt played a massive part in our collective adult success in memorizing not only mathematics, but other important concepts as well.

I may not know why or when "memorizing" became a dirty word in educational terminology, yet it has. Therefore, I have tried to combine my memorization techniques with fun, silly, call-and-response systems that allow students to learn, and administrators to employ them. Throughout my career, I have taught these operations of decimals to students ranging from 2nd grade to senior citizen G.E.D. students. I have tutored fellow teachers in preparation for their licensing exams. What started out as a series of transparencies written in only wet erase markers and transferred carefully to top loading clear projector sheets, has made its way into my teaching vernacular, my permanent jump drives, and now in book and e-book format. It is my sincerest hope and expectation that you will find as much success in my format as I have, and that your students are able to associate their math classes with learning *and* fun.

Sincerely,

Ms. Lisa N. Hernandez

Section Layout

This manual has been divided into the following sections:

- The Script[*]
- The Notes
- The Worksheets
- The Answer Key
- The Appendix

Though, of course, you are free to take from it whenever and wherever you can, I have found my greatest successes with teaching decimals when the lessons are implemented directly in the order given in the book.

---

[*] *Teacher notes in the script have been delineated with italic font.*

Separate Sheet Sideways

A Word about Using a Separate Sheet Sideways...
Or
SSS

OK, I know that you might be thinking that perhaps I have taken the alliteration thing a little too far considering the title of this book. However, this is a very important page. In fact, the basis for this page has been in existence long before I had ever compiled the rest of these lessons. It's simply that I found very quickly, whether using textbooks, workbooks or my own lessons, that too frequently did students arrive at the wrong answer only because their work in the algorithm section[*] was not aligned properly. Taking a regular sheet of notebook paper, and simply turning it sideways, will eliminate virtually all of these mistakes, while simultaneously enforcing student's concepts of place value as well.

I cannot suggest strongly enough that students begin each one of these lessons using a Separate Sheet Sideways (SSS for short). When or if they are able to graduate to regular paper, should be a non-issue. Quite often have I been able to greatly increase the accuracy even of adult learners' when dealing with decimal operations, solely by rotating their notebooks 90 degrees.

---

[*] i.e., they may know exactly what do in the problem, how to do it, and in what order, but may still end up with the wrong answers due solely to the alignment of their work.

Catchphrase Preview

If your students are unfamiliar with the concept of a "catchphrase," use Appendix Page 1 to get them started. Explain that we are going to learn how to calculate decimal operations using quick and easy catchphrases.

The following catchphrases are best remembered by using the following associations:

"Line 'em up" – Cowboy (or military drill sergeant) accent

"Tuck it in" – Glade Plug In commercial

"Kick it Out" – The song and dance "Walk It Out" by Unk

"Bring it up and fuggetaboutit!" – Brooklyn accent and mannerisms

In answer to the question you must be asking of yourself by now, yes, I am aware that I am not the funniest person alive. I'm probably not even in the running for the "funniest teacher writing a book on decimal operations award," for that matter. Too many times, I have been keenly aware that my students are laughing at me, and not with me. Regardless, they remember my silly antics. Especially at state testing time. So there.

That being said, should a funnier or more culturally relevant phrase pop into your head, feel free to go with it. It is the repetition combined with entertainment that is important here.

The Script

Day 1
What Are Decimals?

There are all sorts of fractions in this world. Just for starters, lets name a few off the top of our heads....

Students will call out various fractions and teacher will make a brainstormed list. Keep going until you have a good number of different fractions, leading the students, if necessary, to name some (but not all) fractions with denominators of 10's, 100's, and 1000's.

However, only those fractions with a denominator of 10, 100, 1,000, 10,000, 100,000 (you get the point) are true decimals. Let's circle the true decimals above.[*]

True, decimals are just like abbreviations. You know what abbreviations are, right? Let's think of some together...

Mrs. = _____   USA = _____   St. = _____

Feel free to continue the above vein for as long as necessary or desired to get your point across.

Well decimals are abbreviations as well. They are the abbreviations for any fraction with a denominator of 10, 100, 1000, 10,000, etc!

So let's refresh... What are the abbreviations for $\dfrac{6}{10}$ $\dfrac{7}{10}$ $\dfrac{6}{100}$ $\dfrac{7}{100}$ $\dfrac{6}{1000}$?

Have students call out six tenths, seven tenths, six hundredths, etc. They shouldn't be writing yet, just using words.

Now, the term "abbreviation" is a language arts vocab word, right? So we won't use it any more. We'll use the math term instead, DECIMAL.

Somewhere on the board, overhead, or projection, make a teacher note stating:

Decimal = Abbreviation

Review or quiz throughout as necessary.

---

[*] The others *can* be re-named as decimals, but with a bit of work. We will not worry about these until later. Much later. In another book, in fact!

OK. Now, that we have that down pat. You know about the first part of the number line, right? The one from 1st and 2nd grades? Just in case you need to refresh your memory, here it is

Etc.	Etc.	Thousands	Hundreds	Tens	Ones

But what you may not know is, there is another side to that number line. It looks like this:

Etc.	*Etc.*	Thousands	Hundreds	Tens	Ones	*Tenths*	*Hundredths*	*Thousandths*	*Etc.*	*Etc.*

Just as the left side can go to INFINITY, so can the right side. Just as we know how to name numbers that are on the OLD side of the number line, we are going to learn how to name them on the NEW side. We call the dividing line, the DECIMAL POINT. Clever, huh?

Let's practice with the worksheet *(or a transparency)*. A partner will practice putting a number in a slot to the right of the decimal point, and you will practice saying it. Ready? Let's go!

Partner (or teacher) will practice with transparency or photocopy of Appendix II, placing numerals in different place values, with students naming said place. You may want to use a call-and-response system first, pointing to different place values, then begin filling in different numerals in the chart and students guessing how to correctly name such numbers.

You got that? Now take a look, and go back, and tell me what is the difference between...

- 7 tenths and 7 tens?
- 9 hundreds and 9 hundredths?
- 4 thousandths and 4 thousands?

Which is bigger? How can we tell the difference? If we were talking about money, which would these amounts represent?

If I owed you 40 dollars and I gave you 40 cents, would that be OK? Oh come on, I just had my decimal in the wrong place, right? Certainly not! So for the remainder of this year, this school, middle school, college, university, and your ENTIRE CAREER, if your decimal is off, your answer is INCORRECT. Let's take a minute and make a note of that...

From this day forward, if my decimal is in the <u>wrong place</u>, my answer is <u>wrong</u>.

OK – You can't say that I didn't warn you!

That's why we are going to learn EXACTLY where to put our decimal at all times!

Now, do the opposite of the previous activity. The teacher/partner will call out a number, and the students will write that number correctly using the same decimal chart.

Are you ready to rock? Let's start with the ten**ths**.

| 6 tenths | 2 tenths | 8 tenths | 7 tenths | 1 tenth |

Now, let's do hundred**ths**.

| 4 hundredths | 6 hundredths | 5 hundredths | 3 hundredths |

Now, what about that blank space in our chart to the left of our numeral? What do you think we should put in there? Stars? Bunny rabbits? Our initials? Do we know of any symbols in math that serve as *placeholders*?

Oh yeah, ZEROS! As you can see, you need to use a ZERO as a placeholder to correctly name these decimals. Let's review your answers to see how you did.

We are on a roll now, so let's keep going. This time, we are going to do the thousand**ths**.

| 2 thousandths | 7 thousandths | 9 thousandths | 5 thousandths |

How many ZEROS did you need this time? You get the idea!

Now, we are going to mix it up a little. Let's see who can follow along!

| 2 hundredths | 4 tenths | 3 thousandths | 5 tenths | 7 ten-thousandths |

Now, what if I were to say "9 *and* 4 tenths"? What do you think that might look like? Give it a try. Put what you think that would look like in your chart. Let's check it together.

How many of you put something on paper *(or call a student to the board)* that looked like this?

9.4

If you did, it's clear that you have an amazing Math teacher!

Now, let's try these.

| 4 *and* 6 hundredths | 12 *and* 7 tenths | 9 *and* 2 thousandths |

How did we do? What have you figured out the word "and" means in a decimal problem?

At this point, students should have figured out that verbally we refer to the decimal point as "and." In writing, it looks just like a period in language arts. Prompt them if necessary to put this realization into words.

Now, go ahead and make up three of your own.

_____   _____   _____

Give students an opportunity to quiz their partners (or yourself) with their own examples. Letting them "be the teacher" for a moment or two, this not only lends a good deal of fun to any lesson, but also helps them analyze and synthesize their conceptualization of new information.

Sometimes, a decimal number has more than just one number and a zero, though. Let's take a look at these examples. How do you think we would say them?

.23 .143 .2009 .72

Discuss the correct pronunciations of the decimals above:

23 hundredt*hs*   143 thousandt*hs*   2009 ten-thousandt*hs*   72 hundredt*hs*

Whether through modeling or monitoring their work, make sure the students are grasping the concept of reading the number horizontally (normally) and then tacking on the suffix (vertically) of the final place value in the chart.

As you can see, we read the number as usual, *and then* we add on the name of the place value at the end.

Now, try these:

4.45	10.034	3.001
4 and 45 hundredths	*10 and 34 thousandths*	*3 and 1 thousandths*

16.3	922.14	52.13
16 and 3 tenths	*922 and 14 thousandths*	*52 and 13 hundredths*

Do we have a handle on this? If so, it's time for Worksheet 1. I knew you'd be excited!

Day 2
Comparing and Ordering Decimals

Please refer to page vii regarding the SSS (Separate Sheet Sideways) technique and explain to students what it is. Make sure everyone has one.

Today we are going to work on comparing decimals. Comparing is easy. The best way to compare decimals is to take a regular sheet of lined notebook paper, and turn it sideways. We will do this a lot when dealing with decimals, because it's like an instant place value chart. See what I mean?

A transparency or projection would serve well at this point to familiarize students with using a SSS.

Let's compare the following two decimals: .7 and .5 Which is greater?

That was too easy, right?

But what about this one?   2.07 and .207   How about now?

Which one of these is greater? *Allow students to answer 2.07* Why? I'll tell you why!

Make your notebook paper look exactly like the example in Appendix II. Use a colored pencil or highlighter to mark the entire decimal line up and down on your sideways paper.

This book has a little catchphrase for helping students remember the different procedures for operations:

"Line 'em up"

This is the catchphrase for comparing decimals. *Refer to Appendix I if necessary.*

Let's take out our SSS, shall we? Does everybody have one? Who remembers?

Though clearly this workbook includes reproducible place value charts, it is important that the student be able to see how easy it is to recreate their own with just a blank sheet of loose leaf (SSS) at any time—especially for standardized testing purposes!

Now, put the numbers 2.07 and .207 in your chart, making sure to use the decimal line you just highlighted. Now, is it easier to see which is greater? I thought so. Let's try some even tougher ones! Use the < or > signs to show which is greater.

a) 1.406 ☐ 14.06   b) 300.7 ☐ 30.70   c) 45.64 ☐ .4564

Fantastic! I now dub you "Ready for Worksheet 2A!" Are you flattered?

Allow time for and then discuss answers to Worksheet 2A found in Answer Pages section.

I hope you've been using a SSS to compare these before you decided which was greater. How many of you were? How many feel like you can "just tell" which one is bigger by looking at them?

Well, if you weren't using a SSS before, you are going to now, because you'll definitely need it for a problem like this...

Example 1: List the following decimals in order from least to greatest.

2.456   2.546   2.56   2.405   .2456   2.045

BUT NOT YET. Let's take this little trip down Memory Lane first!

Did you ever alphabetize your spelling words? In 1st grade it was easy. You put all the words that began with "A" before all the words that begin with "C." But what about in 3rd grade? Remember when they began trying to trip us up with words like:

Caterpillar, Cast, Car, Cat, Case

There is a reason I mention this. It's because, just like with decimals, this problem becomes a lot less complicated if you write the words vertically (up and down) first.

Pass out either version, (two separate ones have been provided for you,) of Worksheet 2B.

Caterpillar

Cast

Car

Cat

Case

Obviously, all the words begin with "c," so that's not going to help. In fact, go back and cross out all the C's in the vertical word list. You might as well cross out all the A's for that matter as well. Now, is it becoming a little more obvious which order these words might go in? This is THE SAME THING you need to do with the decimals.

Go back to your SSS and copy the list of decimals (exactly in the order they are written) from Example 1 vertically. Is there anything that jumps right out at us that would automatically be the least in the list?

Great! That's 1 down, 5 to go!

Now, go back and cross out all the 2's. They won't help us decide. Now, there are only two decimals in contention for greatest, right? Circle them, and then cross out the 5's to decide which one is greater. Perfect.

Put that number at the top of your list, then the other one you circled right under it. Now, we have 3 down and only 3 to go. We are making progress!

Next, you have two numbers with 4 and one with 0. Obviously, the one with 0 is going to be less than the other two, right? Put that one next to last. Now, you only have to choose the position of the next two. Cross out both 4's, and see what you have. You have a 5 and a 0. Which one is going to be greater?

Now, we are going to practice this exercise two more times before we go at it independently. Let's try the same (listing these decimals vertically on our SSS). List them in the order they are written, and don't worry about what order they should go in just yet.

.7829   .7892   .8729   .7782   .7987   0.782

Repeat the same process of elimination that we used before, but this time list the numbers from **least to greatest**.

Now, try again with this set, listing from **greatest to least**.

34.567   34.5067   34.5006   34.657   34.506   34.500

Please keep in mind that in one problem, they may ask us to list from **least to greatest**, and in the next, they may ask us to do the exact opposite! Who are these people who write textbooks? Why are they out to make our lives miserable?

Well the answer is.... I have no idea who these evil trolls could be, and why they wish to destroy us. But clearly they do. So you must foil them at every turn by **reading the directions** as to how they want you to list them. (Least to greatest, greatest to least) If you don't... well, then *they* win.

Decimal Fun Fact #1

Did you know that ALL numbers have a decimal? It's true. Even if you don't see it, they all have a decimal. (Psst, it's just behind it, hiding. Don't make any sudden movements! Put your hand out so the decimal can smell it first. OK, maybe I'm going too far). For example, take a look at the following:

7 = 7.   5 = 5.   4 = 4.   210 = 210.

Get the idea?

So now, you should be able to do a list that includes whole numbers, like so...

List the following from least to greatest:

5.1   5.01   5   0.5

Like always, you need to "Line 'Em Up," and fill in any blanks with zeros.

So, hopefully, we are getting good enough now to try this on our own. But first, let's just copy some notes so in case you forget the procedure while you're trying to work independently, you can always refer to them. *See Note Page A.*

Now, we can turn to Worksheet 2C to make sure that we completely understand how to "Compare and Order" decimals.

But first, let's begin our Decimal Notes Chart (Note Page G) and record this latest catchphrase to remind us how to compare decimals.

*Students should create Note Page G, The Complete Decimal Notes Chart, **row by row with each lesson,** and not all at one time.*

Day 3
Adding and Subtracting Decimals

Quite simply...Line 'em up!

You'll remember that this book has a little catchphrase for helping students to remember the different procedures for decimals. "Line 'em up" is not only for comparing, but also the catchphrase for adding and subtracting decimals.

What do we mean? You will, quite literally, take the two numbers that you are trying to add or subtract, get your SSS[*], and line those numbers up **according to the decimal point**. For example...

$$5.2 - 3.1$$

Step 1 — Rewrite it like this — vertically...

$$
\begin{array}{r}
5.2 \\
+3.1 \\
\hline
\end{array}
$$

Step 2 — Bring the decimal point straight down.

$$
\begin{array}{r}
5.2 \\
+3.1 \\
\hline
\underline{}.\underline{}
\end{array}
$$

Step 3 — Then add or subtract as normal.

$$
\begin{array}{r}
5.2 \\
+3.1 \\
\hline
8.3
\end{array}
$$

_____

[*] Remember, for best results, use a SSS (Separate Sheet Sideways)!

Try this one:

$$6.8 + 3.4$$

Be sure to do all three steps!

Now, go back and copy the 3 steps *(or turn to Note page B)* and copy those.

After you finish this, you are ready to do **the first two rows** of Worksheet 3.

*Give students ample time to complete just **the first two rows** of Worksheet 3, and have Answer Page 3 at the ready, as well.*

Let's check our answers from **the first two rows** of Worksheet 3. Did we do well enough to move on?

What if we have a problem that looks like this?

$$4.5 + .678$$

Do we know what to do?

- What about $4 - .37$?
- What about $37.89 + 2.476$?

What do you think that we should do with the empty spaces? What do we usually "hold places with" in math class? Bingo dots? Pieces of cat food? No, silly, ZEROS!

Great! Let's go back, and add this to our notes.

At the bottom of Note Page B *(if they haven't already)* add the following:

Use zeros in any blank spaces to help you add or subtract.

Now, you are ready to finish the rest of Worksheet 3. How exciting!

But first, let's continue our Decimal Notes Chart (Note Page G) and record this latest catchphrase to remind us how to add and subtract decimals.

*Students should continue Note Page G, The Complete Decimal Notes Chart, **row by row with each lesson,** and not all at one time.*

Day 4
Multiplying Decimals

Since we already know how to add and subtract decimals, today we will learn how to multiply them.

Multiplying decimals is easy. There is only a two-step process that you will need to know.

Step 1 – Multiply as usual

Step 2 – "Tuck it in"

That's right. Not until you get your actual answer will you worry about where the decimal goes. Now, what do we mean by "Tuck it in"? Well, you know in math that we always go from right to left when counting numbers. So we will count how many places the decimal is "tucked into" these numbers:

.46　　　.327　　　5.9

In the first example, we see that the decimal is tucked in 2 places. In the second example, it is tucked in 3 places; in the last example, only 1.

When we calculate a decimal multiplication problem, we ADD up the places that the decimals are tucked into in BOTH the first number AND the second that we are multiply by. Let me show you what I mean!

$$\begin{array}{r} .23 \\ \times\ 4.1 \\ \hline \end{array}$$

$$\begin{array}{r} 2.9 \\ \times\ .7 \\ \hline \end{array}$$

19

$$
\begin{array}{r}
.385 \\
\times \quad .43 \\
\hline
\end{array}
$$

$$
\begin{array}{r}
.5 \\
\times \, .7 \\
\hline
\end{array}
$$

By drawing little "humps" under each, make students understand why the answer to each problem respectively is ...

| *3 decimal places in all* | *2 places in all* | *5 places* | *2 places* |

Now, how about we try actually multiplying some decimals AND finding the answers! Get a new SSS and copy these problems down.

A	B	C	D
.23 x 4.1	2.9 x .7	.385 x .43	.5 x .7

So, remember Step 1?

Go back and multiply as usual.

Now, for Step 2. This is fascinating! What's our catchphrase again?

"_____"

In your answer for Example A, **tuck** the decimal point in 3 places. Don't forget to start from the RIGHT.

> For B, tuck it in 2 places.

> For C, tuck it in 5 places.

> For D, just 2 places.

Your answers should come out looking like this:

a) .943   b) 2.03   c) .16555   d) .35

Did they? If so, you are ready for Worksheet 4! Hoorah!

But first, let's go to our Decimal Notes Chart G and record this latest catchphrase to remind us how to multiply decimals.

Day 5
Dividing Decimals

If your students are advanced, whether in age or ability, start here, and skip Day 6 completely. If not, you should at least attempt today's lesson before moving on to Day 6. At any rate, it's your call as the instructor, but they can be used interchangeably.

Have you ever heard the song "Walk It Out?" Well, we're going to KICK it out!

Heck, you may even want to show a "Walk it Out" YouTube clip and have the kids sing and dance to "Kick it Out" as well! This is a nod to kinesthetics and all that jazz.

Again we are going to count places. Hang on, I'll show you what I mean.

You know that a regular division problem looks like this right? For example, $2\overline{)358}$.

We know that the 2 is called the DIVISOR and the 358 is called the DIVIDEND. It is helpful to think of the little $\overline{)}$ sign as a bookend, and then you can remember that the dividend goes inside the bookend, just like its own little shelf!

Decimal Fun Fact #2

Absolutely, positively, under no circumstances may we EVER have decimals in the DIVISOR.

Never in fact.

Then why do book publishers keep giving us problems with the decimal in the DIVISOR? Besides the fact that they may be ornery perhaps? No! To see if we know how to "Kick it out"!

Let's say that our division problem looked like this $.2\overline{)3.58}$

We have no problem with the **3.58**. It's the **.2** that we CANNOT have!

If we could, how many spaces would we have to "kick" this decimal point so that we would have a regular old 2?

One, correct? Just one little place and we could deal with this problem, right?

Well, here is a little secret about division. Lean in, listen closely. We don't want the other classes to hear. Here goes...

YOU CAN DO WHATEVER YOU WANT TO THE DIVISOR as long as you DO THE SAME TO THE DIVIDEND AS WELL!

Shocking, I know. So let's practice a little, with you calling out equal treatments for dividends as for divisors. Let's just say that I wanted to dump cold water on the divisor. What would I have to do to the dividend? What if I wanted to vacuum the divisor? What would I have to do to the dividend? Let's pretend that I pull the divisor's hair? What would we do then? That's what I thought!

So we need to "kick" this decimal out of the DIVISOR one place, so what should we do?

Review preceding paragraph if necessary. Comedy sticks in students' brains and actually makes learning fun! Don't be afraid to repeat as many times as necessary with even more ridiculous examples.

And as long as we do it to the DIVIDEND as well, all is fair.

Let's go back to our example:

$$.2\overline{)3.58}$$

So how many places would we need to "kick out" the decimal in this divisor, in order to have a plain old 2?

If you said, "one place," I am going to build a statue in your honor. Maybe I will call Ripley's because I do NOT believe how smart you are! Thank you for paying such close attention.

So, back to the problem: We said that we need to "kick" this decimal out of the divisor, one place right? So we are allowed to do that if we do the SAME thing in the dividend, remember? So now our problem should look like this...

$$2\overline{)35.8}$$

The key to dividing decimals is we just **bring the decimal point straight up, and forget about it.** Now, we are free to divide like normal human beings, without some pesky decimal bothering us in the DIVISOR *(give divisor dirty look)*.

Let's try some more. How many places would we have to "kick" the decimal out of these problems?

$$\text{a) } 1.7\overline{)3.4}$$

$$\text{b) } .25\overline{)1.75}$$

$$\text{c) } .090\overline{)345.780}$$

$$\text{d) } .003\overline{)9.6}$$

Hopefully you said that in

a) You need to kick it out 1 space, meaning that your new problem would read $17\overline{)34}$

b) In this one, you should have said 2 spaces, turning your problem into $25\overline{)175}$

c) Clearly, this selection required 3 spaces.

d) Hmmm. This one gives us pause for thought. If we "kick the decimal out of the DIVISOR 3X, what do we do with the empty spaces that occur after the 96? Hint: What do we usually do with EMPTY SPACES in math class?

If you said "fill them in with ZEROS," you are correct. If you said "color them green and spray with perfume," you are sorely mistaken.

Now, your problem should look like $3\overline{)9600}$

Much more manageable, wouldn't you say?

Now, GO BACK and actually do the above problems. Now, that you've figured out where the decimal goes, the rest is easy!

Check your answers in the back of the book (Appendix III – Class work for Day 5).

If those are correct, you are ready for Worksheet 5. But first, let's go back to our Decimal Notes Chart G and add this latest catchphrase to remind us how to divide decimals.

If this is NOT completely clear to you, forget about Worksheet 5. In fact, put the book or worksheet down that you are staring at, and walk slowly away. Don't make any sudden movements, because decimals can smell your fear.

OK, that last part is not true. But if you truly did not get dividing it the first time, we are going to try CHUNKING this lesson. *But even if students do understand the above, and you've got some spare time to burn, you may want to give Day 6 lessons a whirl anyway.*

Day 6 (if there is any confusion over day 5) Chunking Division

Chunking Division Part A

OK – remember the fiasco that was yesterday? I need you to totally erase that from your mind. Let's pretend that never happened. Remember that little memory-erasing device from "Men in Black." I need it right now. Anybody have one?

Here is the most important thing about dividing decimals:

Bring it straight up, and fuggedaboutit!

Can I hear you say that a few times? Who can say it loudest, in their best silly accent?

Let's try these problems.

$$3\overline{)13.5} \qquad 2\overline{)97.4} \qquad 10\overline{).64390} \qquad 5\overline{)7.230}$$

What are we going to do with the first problem? That's right, "Bring it straight up, and fuggedaoutit"! What about the second one? You got it, "Bring it straight up, and fuggedaoutit!" But what about the third? Should we "Bring it straight up, and fuggedaoutit?" But now I am completely stumped on the final one. Should I "Line 'em up"? Should I "Tuck it in, tuck it in"? What should I do then? That's right; I should "Bring it straight up, and fuggedaoutit! You sir/ma'am, are a genius!

When completing the problems on board/worksheet/smartboard, I actually encourage students to erase the decimal point in the original problem after bringing it straight up. I'm crazy like that.

Now, it's time to copy Note Page D.

Chunking Division Part B

Use Worksheet 6A and begin again, assigning as many problems as you think necessary at a time. For 6B, rinse and repeat with Day 5 lesson, demonstrating the necessary movement of the decimal point out of the divisor, and the corresponding move in the dividend. In case of emergency, you may have to procure some downloaded Internet worksheets or resort to old-fashioned textbooks (eeew!), depending on the amount of practice with this skill that you feel your students need to obtain mastery. I do, however, try not to spend too much time on any one operation, in fear that they will then forget the rules to the other operations. Now, time for Note Page E!

Chunking Division Part C *(Optional: For use with advanced levels)*

Remember Decimal Fun Fact #1.

Huh? Do ya? Do ya? Did anybody write it down?

OK, I guess I'll have to gently remind you that

ALL NUMBERS HAVE A DECIMAL!

Remember now? Even if you don't see it, they all have a decimal. (Psst, it's just behind it, hiding). For example, take a look at the following:

$$7 = 7. \qquad 5 = 5. \qquad 4 = 4. \qquad 210 = 210.$$

Is it coming back to you?

OK – Fun Fact #2

Now that you have discovered decimals, you will NEVER again use a REMAINDER in any division problem. Decimals and remainders HATE each other. It's true. They refuse to even be in the same room together!

So what do we do instead?

WE USE DECIMALS AND ZEROS!

Hang on for this example:

$$5 \overline{)7}$$

Before this lesson, we might have said that the answer to this problem is 1 r 2, correct? However, now knowing what you do about the long standing rule that decimals and remainders can NEVER be in the same problem, what do we do with that pesky 2 that's hanging around?

I'll show you. Remember Fun Decimal Fact #1? That every number has a decimal whether you see it or not? We are going to use that rule to our advantage. First, let's put a decimal behind that 7.

$$5 \overline{)7.}$$

Once we do that, we are then free to put any number of ZEROS behind that decimal point, which will allow us to continue the problem.

$$5 \overline{)7.0000}$$

Ta Dah!

Let's review:

Step 1 – we wrote the decimal BEHIND the number (it was really already there, we just couldn't see it)

Step 2 – we wrote some ZEROS behind that point

Step 3 – we continued the problem

Let's try those 3 steps with the following problems:

$$2 \overline{)7} \qquad 5 \overline{)8} \qquad 4 \overline{)7} \qquad 8 \overline{)11}$$

But sometimes it seems that there are problems like this that just WON'T GO AWAY! They are like a bad houseguest! You keep working and working and they still are not ready to leave! There are two types of these problems:

First type:

Try these two problems (remember to use the three steps!)

$$3 \overline{)8}$$

$$6 \overline{)8}$$

These are called REPEATING decimals. Can you possibly guess why they might be called that? Mmmhmm. That's what I thought.

Anyway, there is a better way to write repeating decimals. For example, when dividing 12 by 9, rather than writing your answer as

$$1.333333333333333333333333333333333...$$

you can just write $1.\overline{3}$.

See the little line over the 3? That tells us that a decimal is **repeating**.

Second type:

But what about the following problems?

On a **new** SSS (trust me, you'll thank me later), follow the three steps. Go as far as you can with this problem without needing sleep, shelter or refreshments.

$$6 \overline{)7}$$

For those of you who are exceedingly ambitious, give this one a whirl!

$$7 \overline{)8}$$

Is it time for you to graduate yet? Any gray hairs to pluck? Well, after those two problems, or even after just the one, you have to be just a little convinced that textbook publishers make it their daily business to torture school children just as yourselves. And you would be correct; they are. HOWEVER, there is an important CAVEAT (that's a fancy word for "an exception to a rule") that you will benefit from remembering!

Here it goes...

You can usually quit after 3 places to the right of the decimal point. That's right! So after the 1.166 of the first answer, just quit. Feels weird to have a teacher tell you that, huh? I swear it's cool. Go ahead! In the second problem, you could have quit after 1.142. So you are probably smacking yourself in the head saying, "Why didn't she tell us this BEFORE we spent a half hour on each?" Oh, did I forget to tell you? Because secretly, in my other life, I am a textbook publisher, that's why, mwahhahaha! And I hereby order you to continue your Note Page G immediately!

Day 7
Multiplying and Dividing Decimals by 10's, 100's, 1000's, etc.

What do you notice about the following problem? Don't do it. Just tell me if anything jumps out at you...

$$.7 \times 10 = 7$$

What about this one?

$$a)\ .7 \times 100 = 70$$

Or these?

$$b)\ .07 \times 10 = .7 \qquad c)\ .7 \times 1000 = 700$$

Lead students in composing a sentence, whether in writing or out loud, explaining how the answer involves the use of a 7 multiplied by a 1, and the amount of zeros then changes. Only after they figure this out should you show them how many places the decimal point ends up "moving" and in which direction. Have them compare this to the number of zeros in each problem.

That's right! All you have to do when multiplying by 10's, 100's, 1000's, or any other power of ten, is to move the decimal point to the **right**, the **same** number of places as there are zeros. It's so simple, that it's complicated!

And what happens when there are blank spaces such as in examples a) and c)?

That's right, just add zeros!

Who thinks that they could do a whole worksheet like this? *(see Worksheet 7A, first two rows)*
Review answers to Worksheet 7A, first two rows.

OK, now let's switch gears. What do you notice about this problem? Like before, you don't have to do it, let's just make a sentence about it.

$$.4 \div 10 = .04$$

What about this one?

$$a)\ .4 \div 100 = .004$$

29

Or these?

$$\text{b) } .04 \div 10 = .004 \qquad \text{c) } .4 \div 1000 = .0004$$

That's right! All you have to do when **dividing** by 10's, 100's, 1000's, or any other power of ten, is to move the decimal point to the **left**, the **same** number of places as there are zeros. Now, how are we going to remember this?

Here's my little cutesy trick: When a girlfriend and boyfriend break up, would you say that they are multiplying or dividing? Are you familiar with Beyoncé' song, "Irreplaceable"? *Play song clip if necessary. Make sure you get to the "to the left, to the left" part.*

This song is about a couple breaking up, right? She is dividing their things up between what's hers and what's his. They have chosen not to stay together, get married, and multiply (have children), right? I thought so. So every time we see a division sign, we are going to go "to the left, to the left"!

If there is any confusion at this point, you may want to scrap this association and find students in the classroom or famous celebrities with the initials MR or DL. Although I have generally found Beyoncé to be a cultural universal among young people, by the time you are reading this, the song may have lost its relevance.

All right, so to review then, when we see a division sign, we go to the left, the same number of places as there are zeros. And what do we do if there are any blanks? Who did NOT say "fill them in with zeros"? If you did not, I need you to smack yourself in the back of the head.

Let's try the **next two rows** of Worksheet 7A.

Review answers to Worksheet 7A, next two rows.

Alright, I know that you keep saying that you're ready, but I'm just not quite sure. What kind of teacher would I be if I just went around handing out worksheets, willy-nilly? As a professional educator, I need to conserve my worksheets, only to the truly deserving. So who really, really thinks that they are ready? Are you sure? I don't know. I think I'll have to audition you with the last two rows of 7A. If, and only if, you answer the last two rows perfectly, will I bestow upon you the honor of Worksheet 7B.

Review answers to Worksheet 7A, last two rows. Pass out 7B to students only when they have mastered 7A.

I can't believe that this many people are ready for 7B. Did I ever tell you how lucky you were to have such an exemplary Math teacher? I should be famous by now, I swear. It's like I'm running a genius factory here!

Now, we already know how to multiply and divide decimals, whether by powers of 10 or any other number, so we won't include this on Note Page G, since it's not a separate operation. But I will cordially invite you at this point, to copy Note Page F.

Day 8
The Review

Ok, at this point, I actually think any review should be custom built around what you think your students have not mastered. However, it is critical at this time that you review all the catchphrases associated with each operation, lest they become overwhelmed thinking that there forty-three different things to remember. You must remind them (or better yet, they must remind themselves) that there are only four.

See Worksheet 8 review (though a miniature version is right below). Whether using a smartboard, transparency, or giant projection on a whiteboard, point randomly to each boxed problem and encourage students to call out the appropriate catchphrase for each. At this point, they should NOT be worrying about solving the problems, just recognizing and remembering which catchphrase goes with which operation.

Other versions – reproduce said page the size of a bingo card, and call out one of the catchphrases. See how quickly students can "cover" all pertinent problems with bingo chips, pennies, a certain colored crayon mark, etc.

When and only when you are sure that students can immediately recognize HOW to attack decimal operation problems, should you allow them to do so.

Solutions on Answer Page 8

Another idea – Solely from memory, have students recreate Note Page G at this juncture. See if they can do so with no help at all, and then review to fill in any blanks.

.2 + .72	.72 − .2	.2 x .72	$2\overline{)2.72}$	$.2\overline{)2.72}$
$.4\overline{)208}$	.4 − .208	.4 x .208	.4 + .208	$4\overline{)20.8}$
8 + .654	8 − .654	8 x 6.54	$4\overline{).5}$	$.4\overline{)0.5}$
.05 x 2.2	9.2 + .65	8 − .74	$.8\overline{)5}$	0.12 x 1.2
9.0 + 4.93	.62 x 9.1	7 − .97	8 x 5.3	$.6\overline{)7.02}$
$6\overline{)7.02}$	6.5 − .45	20 x .43	$.06\overline{)7.02}$	216 + .9

The Notes

Note Page A

COMPARING DECIMALS

STEP 1: LINE 'EM UP, USING A SEPARATE SHEET SIDEWAYS (SSS)

STEP 2: FILL IN ANY BLANKS WITH ZEROS

STEP 3: "ALPHABETIZE" THEM

EXAMPLE: LIST THE FOLLOWING FROM <u>GREATEST TO LEAST</u>:

8.6   8.06   8   8.006   8.601

8.600
8.060
8.000
8.006
8.601

ANSWER: 8.601,   8.600,   8.060,   8.006

Note Page B/C

ADDING AND SUBTRACTING DECIMALS
"LINE 'EM UP!"

STEP 1 – MAKE SURE THAT YOUR PROBLEM IS WRITTEN TOP TO BOTTOM (VERTICALLY), <u>LINING UP THE DECIMAL POINTS</u> EXACTLY.

EXAMPLE: 4.5 + 6.23

```
   4.50
+  6.23
```

STEP 2 – BRING THE DECIMAL POINT STRAIGHT DOWN.

```
   4.50
+  6.23
     .
```

STEP 3 – ADD OR SUBTRACT LIKE NORMAL!

```
   4.50
+  6.23
  10.73
```

MULTIPLYING DECIMALS
"TUCK IT IN"

STEP 1 – PRETEND THAT THERE ARE NO DECIMAL POINTS IN YOUR PROBLEM. MULTIPLY AS USUAL.

```
  2.3
x 5
 115
```

STEP 2 – COUNT HOW MANY PLACES THE DECIMAL POINTS ARE "TUCKED" INTO THE PROBLEM (BEGINNING FROM THE RIGHT)

```
  2.3
x 5
 115
```

STEP 3 – TUCK YOUR DECIMAL POINT INTO YOUR ANSWER, THAT SAME NUMBER OF PLACES (REMEMBER, BEGINNING FROM THE <u>RIGHT</u>!)

```
  2.3
x 5
 11.5
```

Note Page D

Dividing Decimals

If there is NO decimal in the divisor...

Step 1 – Bring it straight up

Step 2 – and *fuggedaboutit!*

$$2\overline{)12.4}$$

Answer

$$2\overline{)12.4}^{\;6.1}$$

Note Page E

Dividing Decimals

If There IS a Decimal in the Divisor...

Step 1 - Kick it Out

$$.2\overline{)12.4}$$

Step 2 - Then... Bring it straight up and fuggedaboutit!

$$2\overline{)124.}62.$$

Note Page F

Multiplying and Dividing Decimals by Powers of 10

Powers of 10 are

10, 100, 1,000, 10,000, 100,000

and so on and so on until infinity

If you are MULTIPLYING a decimal by any power of ten, **count the zeros**, and move the decimal that many places to the RIGHT.

Example: $25.46 \times 10 = 254.6$ One 0 – <u>one</u> place to the RIGHT

$25.46 \times 100 = 2546$ Two 0's – <u>two</u> places to the RIGHT

$25.46 \times 1000 = 25460$ Three 0's – <u>three</u> places to the RIGHT

If you are DIVIDING a decimal by any power of ten, **count the zeros**, and move the decimal that many places to the LEFT.

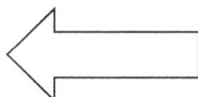

Example: $25.46 \div 10 = 2.546$ One 0 – <u>one</u> place to the LEFT

$25.46 \div 100 = .2546$ Two 0's – <u>two</u> places to the LEFT

$25.46 \div 1000 = .02546$ Three 0's – <u>three</u> places to the LEFT

<u>Three Things To Remember</u>

Multiply – RIGHT   Divide – LEFT
Fill in any blanks with ZEROS

Note Page G

Complete Decimal Notes Chart

Operation	Steps to Solve
Not really an operation, but... **COMPARING**	1. LINE 'EM UP, SSS 2. FILL IN ANY BLANKS WITH ZEROS 3. ALPHABETIZE THEM
ADDING AND SUBTRACTING	1. LINE 'EM UP 2. BRING THE DECIMAL STRAIGHT DOWN 3. ADD OR SUBTRACT LIKE NORMAL
Multiplying	1. Multiply problem as usual. 2. "Tuck" the decimal point in the answer, the same number of places as the problem.
Dividing (divisor only) **Divisor (AND Dividend)**	1. Bring it up & fuggedaboutit! OR 1. Kick it Out then... Bring it up & fuggedaboutit!

The Worksheets

Naming Decimals – Worksheet 1

Using numbers and decimal points, write the following:

a) seven tenths

_____

b) four hundredths

_____

c) twenty one hundredths

_____

d) two thousandths

_____

e) seven hundredths

_____

f) fifteen thousandths

_____

g) five hundred thousandths

_____

h) four thousandths

_____

i) thirty five hundredths

_____

Using word form, write the following:

j) .2 _____

k) .02 _____

l) .002 _____

m) .0002 _____

n) .06 _____

m) .004 _____

n) .62 _____

o) .731 _____

p) .80 _____

q) .03 _____

r) .602 _____

* Hint – Use place value chart to help you.
** Remember that decimal numbers are read horizontally first, then tack on the vertical suffix.
*** Remember, fill in any blanks with _____!!!

Comparing Decimals – Easy – Worksheet 2A

Compare the following decimals using a >, <, or = sign.

	A	B	C	D
1)	2.0 □ 0.2	5.6 □ .05	.72 □ 720	04. □ 4.0
2)	50.5 □ 5.05	.325 □ .0325	7.25 □ 7.521	.5078 □ .598
3)	102.1 □ 102.0	1.7 □ 1.700	8.502 □ 8.520	106.4 □ 104.6
4)	3.5 □ .35	.997 □ 9.07	.26 □ 2.6	.5 □ .500
5)	6.4 □ .64	.92 □ .092	25.99 □ 26	07.03 □ 7.03
6)	4.526 □ 4.256	08. □ 8	10052.0 □ 10051.9	802.34 □ 803.899

Remember to use a SSS!

43

Comparing Decimals – Worksheet 2B, Version I

Remember Our Catchphrase? _____
Remember to fill in any blanks with_____!

List the following words in **alphabetical order.**	Caterpillar Cast Car Cat Case	_____ _____ _____ _____ _____
List the following decimals in order from **greatest to least.**	34.567 34.5067 34.5006 34.657 34.506 34.500	_____ _____ _____ _____ _____ _____
List the following decimals in order from **least to greatest.**	2.456 2.546 2.56 2.405 .2456 2.045	_____ _____ _____ _____ _____ _____
List the following decimals in order from **least to greatest.**	.7829 .7892 .8729 .7782 .7987 0.782	_____ _____ _____ _____ _____
List the following decimals in order from **greatest to least.** Hint – this is a trick question!	5.01 5.1 5 0.5 5.0	_____ _____ _____ _____ _____

Comparing Decimals – Advanced – Worksheet 2B, Version II

Remember Our Catchphrase? _____

Wksht 2B Version 1

.7829
.7892
.8729
.7782
.7987
0.782

List the following decimals in order from least to greatest.

5.1
5.01
5
0.5

List the following decimals in order from least to greatest.

2.456
2.546
2.56
2.405
.2456
2.045

List the following decimals in order from least to greatest.

Caterpillar
Cast
Car
Cat
Case

List the following words in alphabetical order.

34.567
34.5067
34.5006
34.657
34.506
34.500

List the following decimals in order from greatest to least.

Comparing and Ordering Decimals – Worksheet 2C

Catchphrase _____ Remember to fill in any blanks with _____!

Put the following words in alphabetical order:	Put the following decimals in order from greatest to least.	Put the following decimals in order from least to greatest.
Book 1) _____ Battery 2) _____ Boar 3) _____ Bonehead 4) _____ Baton 5) _____ Boondoggle 6) _____	1.02 1) _____ .102 2) _____ 10.2 3) _____	.503 1) _____ .53 2) _____ 50.3 3) _____
Put the following decimals in order from least to greatest.	**Put the following decimals in order from least to greatest.**	**Put the following decimals in order from least to greatest.**
.2 1) _____ .03 2) _____ 1.02 3) _____ 4.002 4) _____ 1.1 5) _____ 1.2 6) _____	.2 1) _____ .02 2) _____ 2.02 3) _____ 2.002 4) _____ 1.2 5) _____ 1.02 6) _____	1.2 1) _____ 3.4 2) _____ 1.02 3) _____ 30.4 4) _____ 10.2 5) _____ 3.04 6) _____
Put the following decimals in order from greatest to least.	**Put the following decimals in order from greatest to least.**	**Put the following decimals in order from least to greatest.**
.009 1) _____ .09 2) _____ 0.09 3) _____ 9 4) _____ 9.09 5) _____ 9.1 6) _____	100.023 1) _____ 100.23 2) _____ 10.203 3) _____ 100.32 4) _____ 1.2030 5) _____ 1000 6) _____	85.03 1) _____ 805.03 2) _____ 80.503 3) _____ 80.53 4) _____ .8053 5) _____ 8.503 6) _____
Put the following decimals in order from greatest to least.	**Put the following decimals in order from greatest to least.**	**Put the following decimals in order from least to greatest.**
07.3 1) _____ .073 2) _____ 7.03 3) _____ .0073 4) _____ 70.3 5) _____ 700.3 6) _____	79.68 1) _____ 7.968 2) _____ .07968 3) _____ .007 4) _____ 709.68 5) _____ 70.968 6) _____	463.7 1) _____ 4.637 2) _____ 46.37 3) _____ 406.37 4) _____ 46.307 5) _____ .4637 6) _____

* Remember to use a place value chart, or better yet, a SSS in order to line up your decimals.
** Also, fill in any blanks with _____.

Adding and Subtracting Decimals – Worksheet 3

Remember Our Catchphrase? _____

7.5 + .9	.8 + 0.4	.6 – .2	1.5 – .6	2.6 + .8
5.4 – .23	8.04 + .7	.9 + 4.6	.78 – .2	6.5 + .9
15.7 + 22.06	56.9 – .07	.523 + 4.6	70.2 – 1.8	.859 – .25
1.256 + .51	9.52 – .7	.456 + 95	5.602 – 3.8	102.7 + 4.69
2 – .52	123 + 4.56	200.895 – 9.72	6095.7 + .83	9.265 – .7

Remember to fill in any blanks with _____!!!

Multiplying Decimals – Worksheet 4

Remember Our Catchphrase? _____

.2 x 9	3 x .8	.05 x 7	.4 x .3	6 x .9
.15 x .2	.21 x .3	.65 x .01	.26 x 4	5.5 x 2
2.1 x 3.4	.21 x 34	5.6 x .12	.82 x .3	75 x .003
.024 x .18	.013 x .2	.750 x 4.2	7.50 x .42	.222 x .33
9.025 x 4.2	2.09 x .36	17.5 x .0002	.023 x .031	4.206 x 7

Remember to fill in any blanks with _____!!!

Mixed Dividing Decimals – Worksheet 5

Remember Our Catchphrase?

_____ and _____

$3\overline{)\,.63}$	$.4\overline{)\,.56}$	$1.2\overline{)\,.48}$	$11\overline{)\,.55}$	$.3\overline{)\,.54}$
$6\overline{)\,.606}$	$.6\overline{)\,4.80}$	$1.3\overline{)\,3.9}$	$2\overline{)\,.9}$	$.4\overline{)\,.7}$
$8\overline{)\,.72}$	$.9\overline{)\,.72}$	$8\overline{)\,8.64}$	$3\overline{)\,.6426}$	$.9\overline{)\,1.08}$
$6\overline{)\,.69}$	$.7\overline{)\,4.90}$	$1.2\overline{)\,.84}$	$2\overline{)\,.5}$	$.9\overline{)\,.5}$
$3\overline{)\,.2}$	$.9\overline{)\,50}$	$.03\overline{)\,1.9}$	$.04\overline{)\,.563}$	$8\overline{)\,.6}$

*Remember to fill in any blanks with _____ !!!

AND - There can be **no remainders in a division problem with decimals!

Dividing Decimals 6A
Remember Our Catchphrase?
_____

$2\overline{)1.4}$	$3\overline{)2.7}$	$5\overline{).45}$	$8\overline{)1.04}$	$11\overline{)4.4}$
$3\overline{)4.5}$	$4\overline{).68}$	$2\overline{)50.2}$	$8\overline{)8.604}$	$6\overline{).6426}$

Dividing Decimals 6B
Remember Our Catchphrase?
_____ and _____

$.2\overline{)6}$	$.7\overline{).49}$	$.3\overline{).201}$	$1.2\overline{)48}$	$.4\overline{)3}$
$8\overline{).6}$	$.2\overline{)1.5}$	$.9\overline{).5}$	$.03\overline{)1.9}$	$.04\overline{).007}$

Remember, fill in any blanks with _____!
AND – There can be no remainders in a division problem with decimals.

Multiplying and Dividing Decimals by 10's, 100's, and 1,000's – Worksheet 7A

Rule: Multiply _____

a	b	c	d
.748 x 10 _____	1.257 x 10 _____	5.73 x 10 _____	1.34 x 10
.748 x 100	1.257 x 100	5.73 x 100	1.34 x 100
.748 x 1000	1.257 x 1000	5.73 x 1000	1.34 x 1000

e	f	g	h
5.2 x 10 _____	.8653 x 10 _____	72.654 x 10 _____	657.2 x 10
5.2 x 100	.8653 x 100	72.654 x 100	657.2 x 100
5.2 x 1000	.8653 x 1000	72.654 x 1000	657.2 x 1000

Rule: Divide _____

a	b	c	d
74.8 ÷ 10 _____	1.257 ÷ 10 _____	57.3 ÷ 10 _____	1.34 ÷ 10
74.8 ÷ 100	1.257 ÷ 100	57.3 ÷ 100	1.34 ÷ 100
74.8 ÷ 1000	1.257 ÷ 1000	57.3 ÷ 1000	1.34 ÷ 1000

e	f	g	h
5.2 ÷ 10 _____	.8653 ÷ 10 _____	72.654 ÷ 10 _____	657.2 ÷ 10
5.2 ÷ 100	.8653 ÷ 100	72.654 ÷ 100	657.2 ÷ 100
5.2 ÷ 1000	.8653 ÷ 1000	72.654 ÷ 1000	657.2 ÷ 1000

Mixed Practice

a	b	c	d
87.24 ÷ 10 _____	684.2 x 10 _____	18.685 x 10 _____	95.7 ÷ 10
87.24 ÷ 100	684.2 x 100	18.685 x 100	95.7 ÷ 100
87.24 ÷ 1000	684.2 x 1000	18.685 x 1000	95.7 ÷ 1000

e	f	g	h
7.6 x 10 _____	101.58 ÷ 10 _____	913.20 ÷ 10 _____	08.10 x 10
7.6 x 100	101.58 ÷ 100	913.20 ÷ 100	08.10 x 100
7.6 x 1000	101.58 ÷ 1000	913.20 ÷ 1000	08.10 x 1000

Mixed Multiplication and Division by 10's, 100's and 1000's – Worksheet 7B

	a	b	c	d	e
1	10 x 5.6	.7 x 100	4.2 x 100	.5 ÷ 10	8.9 x 10
2	1.89 ÷ 100	.5 x 10	0.56 x 10	.89 x 1000	42 ÷ 1000
3	1.89 x 10	7 x 10	.56 ÷ 1000	.7 ÷ 100	1.89 x 100
4	.367 ÷ 10	.5 x 10,000	.367 x 100	0.56 x 10	.07 ÷ 100
5	.34 x 10	.07 ÷ 1000	.42 x 1000	62.05 x 10	.367 ÷ 100
6	36.7 ÷ 10	.07 x 100	367 ÷ 1000	1.89 x 100	5.6 x 100
7	62.05 x 100	62.05 ÷ 10	.089 x 100	.5 x 100	62.05 ÷ 100
8	.07 x 1000	.367 x 1000	1.89 x 1000	5.6 ÷ 100	.042 x 100

Remember, fill in any blanks with _____!

Decimal Operations – Review – **Worksheet 8**

First, Correctly name the catchphrase for each problem.
Then, solve using a SSS.

.2 + .72	.72 − .2	.2 x .72	2)‾2.72	.2)‾2.72
.4)‾.208	.4 − .208	.4 x .208	.4 + .208	4)‾20.8
8 + .654	8 − .654	8 x 6.54	4)‾.5	.4)‾0.5
.05 x 2.2	9.2 + .65	8 − .74	.8)‾5	0.12 x 1.2
9.0 + 4.93	.62 x 9.1	7 − .97	8 x 5.3	.6)‾7.02
6)‾7.02	6.5 − .45	20 x .43	.06)‾7.02	216 + .9

The Answers

Using numbers and decimal points, write the following:

a) seven tenths
.7

b) four hundredths
.04

c) twenty one hundredths
.21

d) two thousandths
.002

e) seven hundredths
.07

f) fifteen thousandths
.015

g) five hundred thousandths
.00005

h) four thousandths
.004

i) thirty five hundredths
.35

Using word form, write the following:

j)   .2   _____two tenths_____

k)   .02   _____two hundredths_____

l)   .002   _____two thousandths_____

m)   .0002   _____two ten thousandths_____

n)   .06   _____six hundredths_____

o)   .004   _____four thousandths_____

p)   .62   _____sixty two hundredths_____

q)   .731   _____seven hundred and thirty one thousandths_____

r)   .80   _____eighty hundredths OR eight tenths_____

s)   .03   _____three hundredths_____

k)   .602   _____six hundred two thousandths_____

* Hint – Use the place value chart to help you.

** Remember that decimal numbers are read horizontally first, then tack on the vertical suffix.

*** Remember, fill in any blanks with _____.

Comparing Decimals – Easy   Answers 2A

Compare the following decimals using the greater than symbol, >, the less than symbol, <, or the equal symbol, =.

	A	B	C	D
1)	2.0 > 0.2	5.6 > .05	.72 < 720	04. = 4.0
2)	50.5 > 5.05	.325 > .0325	7.25 < 7.521	.5078 < .598
3)	102.1 > 102.0	1.7 = 1.700	8.502 < 8.520	106.4 > 104.6
4)	3.5 > .35	.997 < 9.07	.26 < 2.6	.5 = .500
5)	6.4 > .64	.92 > .092	25.99 < 26	07.03 = 7.03
6)	4.526 > 4.256	08. = 8	10052.0 > 10051.9	802.34 < 803.899

Remember to use a SSS!

Answers are circled

ex 1

List the following words in alphabetical order.	Caterpillar ✓
	Cast ✓
	Car ✓
	Cat ✓
	Case ✓

Car
Case
Cast
Cat
Caterpillar

List the following decimals in order from least to greatest.	2.456
	2.546
	2.56
	2.405
	.2456
	2.045

2	.	4	5	6	
2	.	5	4	6	
2	.	5	6		
2	.	4	0	5	
	.	2	4	5	6
2	.	0	4	5	

ex 2

List the following decimals in order from least to greatest.	.7829
	.7892
	.8729
	.7782
	.7987
	0.782

7	8	2	9	✓
7	9	9	2	✓
8	7	2	9	
7	7	8	2	✓
7	9	8	7	
7	8	2		

.7782
.782
.7829
.7892
.7987
.8729

ex 4

.	2	4	5	6
2	.	0	4	5
2	.	4	0	5
2	.	4	5	6
2	.	5	6	
2	.	5	4	6

ex 3

List the following decimals in order from greatest to least.	34.567
	34.5067
	34.5006
	34.657
	34.506
	34.500

3	4	.	6	5	7
3	4	.	5	6	7
3	4	.	5	0	6	7
3	4	.	5	0	6
3	4	.	5	0	0

3	4	.	6	5	7
3	4	.	5	6	7
3	4	.	5	0	6	7
3	4	.	5	0	0	6
3	4	.	5	0	0

List the following decimals in order from least to greatest.	5.1
	5.01
	5
	0.5

Hint: which is greatest

5	.	1
5	.	0	1
5		
0	.	5

	5
5	.	0
5	.	0	1
5	.	1

Comparing Decimals　　　　Answers 2B Version II

Remember Our Catchphrase?
_____ "Line 'Em Up!" _____
Remember to fill in any blanks with **zeros**!

List the following words in **alphabetical order.**	Caterpillar Cast Car Cat Case	Car Case Cast Cat Caterpillar
List the following decimals in order from **greatest to least.**	34.567 34.5067 34.5006 34.657 34.506 34.500	34.657 34.567 34.5067 34.506 34.5006 34.500
List the following decimals in order from **least to greatest.**	2.456 2.546 2.56 2.405 .2456 2.045	.2456 2.045 2.405 2.456 2.546 2.56
List the following decimals in order from **least to greatest.**	.7829 .7892 .8729 .7782 .7987 0.782	.7782 0.782 .7829 .7892 .7987 .8729
List the following decimals in order from **greatest to least.** Hint – this is a trick question!	5.01 5.1 5 0.5 5.0	5.1 5.01 5 5.0 0.5

Comparing and Ordering Decimals   Answers 2C

Catchphrase
____"Line 'Em Up!"____

Put the following words in alphabetical order:	Put the following decimals in order from greatest to least.	Put the following decimals in order from least to greatest.
Book 1) Baton Battery 2) Battery Boar 3) Boar Bonehead 4) Bonehead Baton 5) Book Boondoggle 6) Boondoggle	1.02 1) 10.2 .102 2) 1.02 10.2 3) .102	.503 1) .503 .53 2) .53 50.3 3) 50.3
Put the following decimals in order from least to greatest.	Put the following decimals in order from least to greatest.	Put the following decimals in order from least to greatest.
.2 1) .03 .03 2) .2 1.02 3) 1.02 4.002 4) 1.1 1.1 5) 1.2 1.2 6) 4.002	.2 1) .02 .02 2) .2 2.02 3) 1.02 2.002 4) 1.2 1.2 5) 2.002 1.02 6) 2.02	1.2 1) 30.4 3.4 2) 10.2 1.02 3) 3.4 30.4 4) 3.04 10.2 5) 1.2 3.04 6) 1.02
Put the following decimals in order from greatest to least.	Put the following decimals in order from greatest to least.	Put the following decimals in order from least to greatest.
.009 1) 9.1 .09 2) 9.09 0.09 3) 9 9 4) 0.09 9.09 5) .09 9.1 6) .009 *0.09 and .09 are EQUAL	100.023 1) 1000 100.23 2) 100.32 10.203 3) 100.23 100.32 4) 100.023 1.2030 5) 10.203 1000 6) 1.2030	85.03 1) .8053 805.03 2) 8.503 80.503 3) 80.503 80.53 4) 80.53 .8053 5) 85.03 8.503 6) 805.03
Put the following decimals in order from greatest to least.	Put the following decimals in order from greatest to least.	Put the following decimals in order from least to greatest.
07.3 1) 700.3 .073 2) 70.3 7.03 3) 7.3 .0073 4) 7.03 70.3 5) .073 700.3 6) .0073	79.68 1) 709.68 7.968 2) 79.68 .07968 3) 70.968 .007 4) 7.968 709.68 5) .07968 70.968 6) .007	463.7 1) .4637 4.637 2) 4.673 46.37 3) 46.307 406.37 4) 46.37 46.307 5) 406.37 .4637 6) 463.7

Remember to use a place value chart, or better yet, a SSS in order to line up your decimals.

Adding and Subtracting Decimals   Answers 3

Remember Our Catchphrase?
____Line 'Em Up! ____

7.5 + .9 8.4	.8 + 0.4 1.2	.6 − .2 .4	1.5 − .6 .9	2.6 + .8 3.4
5.4 − .23 5.17	8.04 + .7 8.74	.9 + 4.6 5.5	.78 − .2 .58	6.5 + .9 7.4
15.7 + 22.06 37.76	56.9 − .07 56.83	.523 + 4.6 5.123	70.2 − 1.8 68.4	.859 − .25 .609
1.256 + .51 1.766	9.52 − .7 8.82	.456 + 95 95.456	5.602 − 3.8 1.802	102.7 + 4.69 107.39
2 − .52 1.48	123 + 4.56 127.56	200.895 − 9.72 191.175	6095.7 + .83 6096.53	9.265 − .7 8.565

Remember to fill in any blanks with __zeros__!!!

Multiplying Decimals   Answers 4

Remember Our Catchphrase?
_____ "Tuck It In" _____

.2 x 9	3 x .8	.05 x 7	.4 x .3	6 x .9
1.8	2.4	.35	.12	5.4
.15 x .2	.21 x .3	.65 x .01	.26 x 4	5.5 x 2
.030	.063	.0065	1.04	11.0
2.1 x 3.4	.21 x 34	5.6 x .12	.82 x .3	75 x .003
7.14	7.14	.672	.246	.225
.024 x .18	.013 x .2	.750 x 4.2	7.50 x .42	.222 x .33
.00432	.0026	3.15	3.15	.07326
9.025 x 4.2	2.09 x .36	17.5 x .0002	.023 x .031	4.206 x 7
37.905	.7524	.0035	.000713	29.442

Remember to fill in any blanks with __zeros__!!!

Mixed Dividing Decimals   Answers 5

Remember Our Catchphrase?
_____"Bring It Straight Up___   and   ____Fuggedaboutit!"_____

$3\overline{).63}$ .21	$.4\overline{).56}$ 1.4	$1.2\overline{).48}$ .4	$11\overline{).55}$ .05	$.3\overline{).54}$ 1.8
$6\overline{).606}$ .101	$.6\overline{)4.80}$ 8	$1.3\overline{)3.9}$ 3	$2\overline{).9}$ .45	$.4\overline{).7}$ 1.75
$8\overline{).72}$ .09	$.9\overline{).72}$ .8	$8\overline{)8.64}$ 1.08	$3\overline{).6426}$ .2142	$.9\overline{)1.08}$ 1.2
$6\overline{).69}$ 1.15	$.7\overline{)4.90}$ 7	$1.2\overline{).84}$ .7	$2\overline{).5}$ .25	$.9\overline{).5}$ $.\overline{5}$
$3\overline{)\;.2}$ $.0\overline{6}$	$.9\overline{)\;50}$ $55.\overline{5}$	$.03\overline{)\;1.9}$ $63.\overline{3}$	$.04\overline{).563}$ 14.075	$8\overline{).6}$ .075

Remember to fill in any blanks with zeros!!!

AND – There can be **no remainders** in a division problem with decimals!

Dividing Decimals 6A   Answers 6

Remember Our Catchphrase?
Bring It Straight Up and Fuggedaboutit!

$\dfrac{.7}{2\overline{)1.4}}$	$\dfrac{.9}{3\overline{)2.7}}$	$\dfrac{.09}{5\overline{).45}}$	$\dfrac{.13}{8\overline{)1.04}}$	$\dfrac{.4}{11\overline{)4.4}}$
$\dfrac{1.5}{3\overline{)4.5}}$	$\dfrac{.17}{4\overline{).68}}$	$\dfrac{25.1}{2\overline{)50.2}}$	$\dfrac{1.0755}{8\overline{)8.604}}$	$\dfrac{.1071}{6\overline{).6426}}$
	Dividing Decimals 6B **Remember Our Catchphrase?** Kick It Out___ Bring It Straight Up and Fuggedaboutit!			
$\dfrac{30}{.2\overline{)6}}$	$\dfrac{.7}{.7\overline{).49}}$	$\dfrac{.67}{.3\overline{).201}}$	$\dfrac{40}{1.2\overline{)48.}}$	$\dfrac{7.5}{.4\overline{)3}}$
$\dfrac{.075}{8\overline{).6}}$	$\dfrac{.75}{.2\overline{)1.5}}$	$\dfrac{.\overline{5}}{.9\overline{).5}}$	$\dfrac{63.\overline{3}}{.03\overline{)1.9}}$	$\dfrac{.175}{.04\overline{).007}}$

Remember, fill in any blanks with zeros!

AND – There can be **no remainders** in a division problem with decimals!

Multiplying and Dividing Decimals by 10's, 100's, and 1,000's   Answers 7A

Rule: Multiply__RIGHT							
a		b		c		d	
.748 x 10	7.48	1.257 x 10	12.57	5.73 x 10	57.3	1.34 x 10	13.4
.748 x 100	74.8	1.257 x 100	125.7	5.73 x 100	573	1.34 x 100	134
.748 x 1000	748	1.257x 1000	1257	5.73 x 1000	5730	1.34 x 1000	1340
e		f		g		h	
5.2 x 10	52	.8653 x 10	8.653	72.654 x 10	726.54	657.2 x 10	6572
5.2 x 100	520	.8653 x 100	86.53	72.654 x 100	7265.4	657.2 x 100	65720
5.2 x 1000	5200	.8653 x 1000	865.3	72.654 x 1000	72654	657.2 x 1000	657200

Rule: Divide____LEFT

a		b		c		d	
74.8 ÷10	7.48	1.257 ÷ 10	.1257	57.3 ÷ 10	5.73	1.34 ÷ 10	.134
74.8 ÷ 100	.748	1.257 ÷ 100	.01257	57.3 ÷ 100	.573	1.34 ÷ 100	.0134
74.8 ÷ 1000	.0748	1.257 ÷ 1000	.001257	57.3 ÷1000	.0573	1.34 ÷ 1000	.00134
e		f		g		h	
5.2 ÷ 10	.52	.8653 ÷ 10	.08653	72.654 ÷ 10	7.2654	657.2 ÷ 10	65.72
5.2 ÷ 100	.052	.8653 ÷ 100	.008653	72.654 ÷ 100	.72654	657.2 ÷ 100	6.572
5.2 ÷ 1000	.0052	.8653 ÷ 1000	.0008653	72.654 ÷ 1000	.072654	657.2 ÷ 1000	.6572

Mixed Practice

a		b		c		d	
87.24 ÷ 10	8.724	684.2 x 10	6842	18.685 x 10	186.85	95.7 ÷ 10	9.57
87.24 ÷ 100	.8724	684.2 x 100	68420	18.685 x 100	1868.5	95.7 ÷ 100	.957
87.24 ÷ 1000	.08724	684.2 x 1000	684200	18.685 x 1000	18685	95.7 ÷ 1000	.0957
e		f		g		h	
7.6 x 10	76	101.58 ÷ 10	10.158	913.20 ÷ 10	91.32	08.10 x 10	81
7.6 x 100	760	101.58 ÷ 100	1.0158	913.20 ÷ 100	9.132	08.10 x 100	810
7.6 x 1000	7600	101.58 ÷ 1000	.10158	913.20 ÷ 1000	.9132	08.10 x 1000	8100

Mixed Multiplication and Division by 10's, 100's and 1000's Answers 7B

	a	b	c	d	e
1	10 x 5.6 56	.7 x 100 70	4.2 x 100 420	.5 ÷ 10 .05	8.9 x 10 89
2	1.89 ÷ 100 .0189	.5 x 10 5	0.56 x 10 5.6	.89 x 1000 890	42 ÷ 1000 .042
3	1.89 x 10 18.9	7 x 10 70	.56 ÷1000 .00056	.7 ÷ 100 .007	1.89 x 100 189
4	.367 ÷ 10 .0367	.5 x 10,000 5000	.367 x 100 36.7	0.56 x 10 5.6	.07 ÷ 100 .0007
5	.34 x 10 3.4	.07 ÷ 1000 .00007	.42 x 1000 420	62.05 x 10 620.5	.367 ÷ 100 .00367
6	36.7 ÷ 10 3.67	.07 x 100 7	367 ÷ 1000 .367	1.89 x 100 189	5.6 x 100 560
7	62.05 x 100 6205	62.05 ÷ 10 6.205	.089 x 100 8.9	.5 x 100 50	62.05 ÷ 100 .6205
8	.07 x 1000 70	.367 x 1000 367	1.89 x 1000 1890	5.6 ÷ 100 .056	.042 x 100 4.2

Remember, fill in any blanks with zeros!

65

Decimal Operations – Review   Answers 8

First – Correctly name the catchphrase for each problem. Then – solve using a SSS.

.2 + .72 .92	.72 – .2 .52	.2 x .72 .144	$2\overline{)2.72}$ with 1.36	$.2\overline{)2.72}$ with 13.6
$.4\overline{).208}$ with $.52$	.4 – .208 .192	.4 x .208 .0832	.4 + .208 .608	$4\overline{)20.8}$ with 5.2
8 + .654 8.654	8 – .654 7.346	8 x 6.54 52.32	$4\overline{).5}$ with $.125$	$.4\overline{)0.5}$ with 1.25
.05 x 2.2 .11	9.2 + .65 9.85	8 – .74 7.26	$.8\overline{)5}$ with 6.25	0.12 x 1.2 .144
9.0 + 4.93 13.93	.62 x 9.1 5.642	7 – .97 6.03	8 x 5.3 42.4	$.6\overline{)7.02}$ with 11.7
$6\overline{)7.02}$ with 1.17	6.5 – .45 6.05	20 x .43 8.6	$.06\overline{)7.02}$ with 117	216 + .9 216.9

66

The Appendix

Appendix I

Catchphrase Worksheet

(Naturally, this exercise can be enhanced by any Internet or audio back-up possible)

<u>Famous Catchphrases and Their Originators</u>

In case students are unfamiliar with the term "catchphrase," perhaps some of those below will help define the idea. You may want to have them guess one column or the other, or have them use the blank spaces to make up some of their own pop-culture examples of catchphrases.

Catchphrase	Character
Bart Simpson	"Aye caramba!"
Bugs Bunny	"Eh, what's up, Doc?"
Mr. T from the A-Team	"I Pity the Fool"
Tony the Tiger	"They're grrrrreat!"
Mr. Rogers	"Won't you be my neighbor?"
Shaggy from Scooby Doo	"Zoinks!"
Donald Trump	You're Fired
Erkel	Did I do that?

Appendix II

Place Value Chart

etc.	
etc.	
millionths	
hundred thousandths	
ten thousandths	
thousandths	
hundredths	
tenths	
ones	
tens	
hundreds	
thousands	
ten thousands	
hundred thousands	
millions	
ten millions	
hundred millions	
etc.	
etc.	

Classwork Answers for Day 5

a) $25\overline{)1.75}$ ... -175 ... 0 ... 7

b)

c)

d)

The TESTS

Test 1

Name_____

Fill in each blank in the 1st column with the appropriate answer from the 2nd column.

1. Decimals are just _____ for fractions that have a denominator of 10, 100, or 1,000, etc.	a. zeros
2. Just as on the left side of the number line, the right side of it, after the decimal, also goes on for _____.	b. "Line 'Em Up!"
3. Use _____ as placeholders in any decimal operations.	c. "Bring it straight up and "fuggedaboutit!"
4. If your decimal point is even one digit off, your answer is still _____.	d. abbreviations
5. This is the catchphrase for adding or subtracting decimals _____.	e. remainders
6. This is the catchphrase for multiplying decimals _____.	f. incorrect
7. This is the catchphrase for dividing decimals _____	g. "Tuck It In"
8. There can be no _____ in a decimal problem.	h. infinity
9. Using a SSS is helpful because we can use the lines as a _____.	i. right
10. All numbers actually have a _____, whether we can see it or not.	j. comparing decimals
11. When multiplying decimals by any power of 10's, (100's, 1,000's, etc.) just move the decimal point the same amount of zeros, to the _____.	k. left
12. When dividing decimals by any power of 10's, (100's, 1,000's, etc.) just move the decimal point the same amount of zeros, to the _____.	l. place value chart
13. Though not really a decimal operation, _____ uses the same catchphrase as addition and subtraction of decimals.	m. decimal point

Test 2

Decimal Operations

First, Correctly name the catchphrase for each problem.
Then, solve using a SSS.

.4 + .76	.93 − .9	.1 x .520	$4\overline{)2.72}$	$.2\overline{)4.5}$
$.6\overline{).321}$	8.4 − .208	.5 x .46	3 + .954	$3\overline{)21.3}$
.6 + 1.245	11 − .23	7 x 1.51	$3\overline{).4}$	$.4\overline{)0.6}$
.21 x 3.2	5.2 + .15	19 − .74	$.4\overline{)2}$	0.51 x 1.5
170 + 3.91	.72 x 4.1	5 − .67	6 x 3.1	$.4\overline{)5.28}$
$9\overline{)6.18}$	2.5 − .3	15 x .62	$.05\overline{)11.02}$	.548 + .9

The TESTS Answers

Test 1

Fill in each blank in the 1st column with the appropriate answer from the 2nd column.

1. Decimals are just _____**d - abbreviations**_____ for fractions that have a denominator of 10, 100, or 1,000, etc.	a. zeros
2. Just as on the left side of the number line, the right side of it, after the decimal, also goes on for _____**h - infinity**_____.	b. "Line 'Em Up!"
3. Use _____**a - zeros**_____ as placeholders in any decimal operations.	c. "Bring it straight up and "fuggedaboutit!"
4. If your decimal point is even one digit off, your answer is still _____**f - incorrect**_____.	d. abbreviations
5. This is the catchphrase for adding or subtracting decimals _____**b - "Line 'Em Up"**_____.	e. remainders
6. This is the catchphrase for multiplying decimals _____**g - "Tuck It In"**_____.	f. incorrect
7. This is the catchphrase for dividing decimals _**c - Bring it straight up and fuggedaboutit!"**_.	g. "Tuck It In"
8. There can be no _____**e - remainders**_____ in a decimal problem.	h. infinity
9. Using a SSS is helpful because we can use the lines as a _____**l - place value chart**_____.	i. right
10. All numbers actually have a _____**m - decimal point**_____, whether we can see it or not.	j. comparing decimals
11. When multiplying decimals by any power of 10's, (100's, 1,000's, etc.) just move the decimal point the same amount of zeros, to the _**i - right**_.	k. left
12. When dividing decimals by any power of 10's, (100's, 1,000's, etc.) just move the decimal point the same amount of zeros, to the _**k - left**_.	l. place value chart
13. This action, though not really a decimal operation, _____**j - comparing decimals**_____ uses the same catchphrase as addition of decimals.	m. decimal point

73

Test 2

Decimal Operations (answers)

First, Correctly name the catchphrase for each problem.
Then, solve using a SSS.

.4 + .76 1.16	.93 − .9 .03	.1 x .520 .052	$4\overline{)2.72}$ .68	$.2\overline{)4.5}$ 22.5
$.6\overline{).321}$ .535	8.4 − .208 8.192	.5 x .46 .23	3 + .954 3.954	$3\overline{)21.3}$ 7.1
.6 + 1.245 1.845	11 − .23 10.77	7 x 1.51 10.57	$3\overline{).4}$ $.1\bar{3}$	$.4\overline{)0.6}$ 1.5
.21 x 3.2 .672	5.2 + .15 5.35	19 − .74 18.26	$.4\overline{)2}$ 5	0.51 x 1.5 .765
170 + 3.91 173.91	.72 x 4.1 2.952	5 − .67 4.33	6 x 3.1 18.6	$.4\overline{)5.28}$ 13.2
$9\overline{)6.18}$ $.68\bar{6}$	2.5 − .3 2.2	15 x .62 9.3	$.05\overline{)11.02}$ 220.4	.548 + .9 1.448